MW00895903

IRAQ

Major World Nations
IRAQ

J.P. Docherty

CHELSEA HOUSE PUBLISHERS
Philadelphia

Chelsea House Publishers

Copyright © 1999 by Chelsea House Publishers,
a division of Main Line Book Co.
All rights reserved.
Printed in Hong Kong

First Printing.

1 3 5 7 9 8 6 4 2

Library of Congress Cataloging-in-Publication Data

Docherty, J.P.
Iraq / J.P. Docherty.
p. cm. — (Major world nations)
Includes index.
Summary: Explores the people, history, culture, land, climate, and economy of Iraq, the
"Cradle of Civilization."
ISBN 0-7910-4979-5 (hc)
1. Iraq—Juvenile literature. [1. Iraq.] I. Title.
II. Series.
DS70.6.D63 1998
956.7—dc21   98-4310
CIP
AC

ACKNOWLEDGEMENTS

The author and publishers are grateful to the following organizations and individuals for
permission to reproduce copyright illustrations in this book:
Hutchison Photo Library; The Mansell Collection Ltd.; Nature Photographers Ltd.;
Popperfoto; Frank Spooner Pictures Ltd.

4

CONTENTS

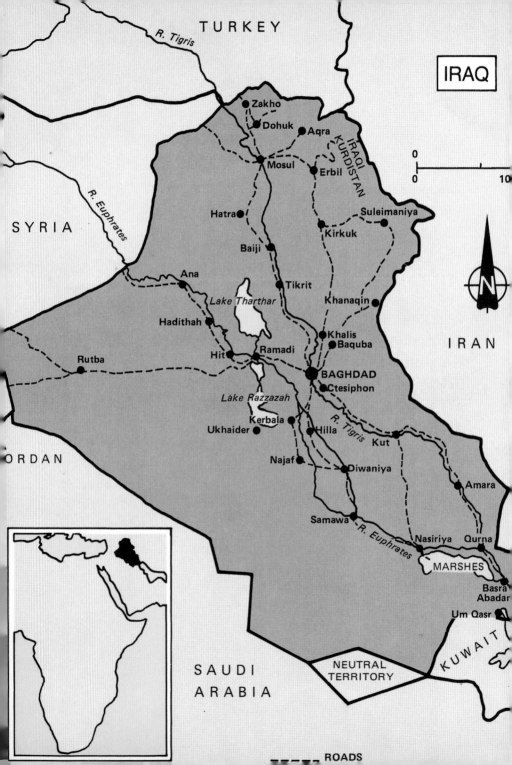

FACTS AT A GLANCE

Land and People

Official Name	Republic of Iraq
Location	In the Middle East on the northwest coast of the Persian Gulf, west of Iran and northeast of Saudi Arabia
Area	167,875 square miles (435,052 square kilometers)
Climate	Dry desert climate
Capital	Baghdad
Other Cities	Mosul, Erbil, Kirkuk, Basra
Population	22,219,000
Population Density	132.3 people per square mile (51.1 people per square kilometer)
Major Rivers	Tigris, Euphrates
Official Language	Arabic
Religions	Shi'a Islam (50 percent); Sunni Muslim (45 percent); Christian (2 percent)
Literacy Rate	58 percent

Average Life Expectancy	57.3 years for males; 60.4 years for females

Economy

Natural Resources	Oil, natural gas
Agricultural Products	Wheat, tomatoes, dates, livestock
Other Products	Oil, petroleum products, phosphates
Industries	Petroleum, farming, manufacturing, transportation
Major Imports	Agricultural products, manufactured goods
Major Exports	Oil, dates
Currency	Iraqi dinar

Government

Form of Government	Constitutional republic
Government Bodies	National Assembly; Revolutionary Command Council
Formal Head of State	President

HISTORY AT A GLANCE

4000 B.C.   The city of Sumer is built at the mouth of the Euphrates on the Persian Gulf. It is the model for all the later city-states of the Middle East and Mediterranean world.

3000 B.C.   Centered around their original city-state of Ashur in northern Iraq, the Assyrians first emerge as a regional power.

2249 B.C.   The Akkadians, a Semitic people, conquer the Sumerians. Although their empire does not last, the Akkadian language is widely used for diplomatic and business communications in the ancient Middle East.

1700 B.C.   Under Hammurabi, the city-state of Babylon (near modern Baghdad) first comes to prominence. Hammurabi writes the first legal code.

1116 B.C.   The Assyrian empire revives and reaches its greatest extent.

616 B.C.   Assyrian power finally ends and a new Babylonian Empire rises.

538 B.C.   Babylon is conquered by the Persian Empire,

which was based on the territory now known as Iran.

235 B.C. Iraq becomes part of the Parthian Empire.

661 A.D. The Islamic world falls under the leadership of the Umayyad Dynasty. Their capital and center of interest is in Damascus, but they control most of Iraq.

750 A new dynasty, the Abbasids, with its origins among the original Muslims of Arabia, overthrows the Umayyads.

763 Baghdad is founded as the new capital of the Abbasid Dynasty.

1055 Seljuk Turks conquer Baghdad.

1258 The Mongols destroy Baghdad, ending the great era of medieval Islamic culture.

1533 Ottomans complete conquest of northern Iraq.

1547 Southern Iraq added to Ottoman Empire.

1917 British forces occupy Baghdad (March 11) after encountering several years of stiff resistance by Turkish armies on the Tigris and Euphrates Rivers.

1918 British troops drive remaining Turkish troops from Iraq, taking possession of the oil fields around Mosul.

1941 Iraqi nationalists start a revolt against Britain. The revolt is stopped without undue force, but most of the leaders are later hanged.

1955 Iran, Iraq, Pakistan, Turkey and Great Britain

form a mutual defense organization called the Baghdad Pact.

1958 King Faisal is killed in a coup (July 14) and the Hashemite kingdom is replaced by a military government. Abdel Karim Kassem becomes the new prime minister.

1959 Iraq withdraws from the Baghdad Pact.

1961 The Kurdish minority, concentrated in northwestern Iraq start a revolt.

1963 Members of the Baathist party come to power.

1970 Iraq nationalizes oil deposits, most of which are owned by British companies. Britain attempts to freeze Iraqi bank deposits in London, but finds thay have been moved to Switzerland. Saddam Hussein is a key player in both operations.

1973 During the Yom Kippur War, Iraq comes to the aid of Syria on the Golan Heights.

1979 Saddam Hussein emerges as the dominant figure among the Baathist leadership.

1980-1988 Iraq fights a long war with Iran over disputed oil-rich territory. Iraq's Kurdish minority starts a revolt and is brutally crushed. Iraq incurs large debts with Saudi Arabia and Kuwait for armaments money.

1990 Disputes over war debts and oil deposits lead to an Iraqi occupation of Kuwait (August 4). An international coalition forms to oppose Iraq's actions.

1991 Iraq is devastated by a massive military attack by an international coalition (January and February).

Kuwait is liberated and some outlying Kurdish and Shi'ite areas of Iraq are occupied by Western troops.

1990s Saddam Hussein and the Baathist regime remain in power. An international embargo makes food and medicine scarce. Iraq frequently interferes with UN inspectors looking for illegal weapons, but international support for another military attack is lacking.

1

Iraq: The Cradle of Civilization

Iraq is a modern Arab state lying at the head of the Arabian Gulf and encircled by Kuwait, Saudi Arabia, Jordan, Syria, Turkey and Iran. It has an area of 168,000 square miles (435,000 square kilometers) and has a population of over twenty million. Iraq is a country of contrasting climates, landscapes and peoples and, not least, the legacies of over five thousand years of history.

Iraq is a land of desert, mountain, marsh and river. These geographical factors have shaped its history from the Sumerian period (4000-2000 B.C.) to the present day. Much of the country is a vast alluvial plain (its fertile soil the remains of deposits from the flooding rivers), baking in summer temperatures of 110 to 120 degrees Fahrenheit (45 to 50 degrees Celsius). For thousands of years great cities were built by powerful rulers along the flood plains of Iraq's lifelines—her two rivers, the Tigris (*Al Dijlis*) and the Euphrates (*Al Furaat*)—and across the apex of the so-called Fertile Crescent. Mesopotamia—"the land between the rivers"—was the name given to this country by the ancient Greeks. Throughout history, kingdoms and city-states flourished here, trading on the

13

A view of the Euphrates River—for centuries this river has been a vital trading route and lifeline for the people of Iraq.

great river routes through Basra to the Arabian Gulf and thence to India and the Orient or East Africa, and on the east-west overland caravan routes from Ancient Egypt, Carthage and the Syrian kingdoms through southern Anatolia (Turkey) and Mesopotamia to Persia (Iran) and beyond.

The ancient lands of Sumer and Akkad were the home of the Sumerians three thousand years before Christ. Archaeological evidence has shown us that it was in this area and at around this time that the earliest cultures in western Asia and perhaps in history were developing. For this reason Mesopotamia is sometimes referred to as the "Cradle of Civilization." In these kingdoms the roots of language, writing, literature, laws, agriculture, science, mathematics,

14

planning, administration and scholarship were established.

With few natural barriers, Mesopotamia has always been open to invaders. The Akkadian and Sumerian city-states were succeeded by the Assyrian and Babylonian empires; Babylon fell to the Persians in the sixth century B.C. and the country remained part of the Persian Empire for two hundred years, until Alexander the Great became the first European (Greek) conqueror of the region. The Greek Empire was in turn replaced by the Persians, who ruled from about 150 B.C. to the advent of Islam, and the foundation of Baghdad by the conquering Arabs in the seventh century. Under the Arab Abbasid dynasty Baghdad flourished as a center of Islamic culture until it fell to Mongol invaders in 1258.

Until this time Iraq had a thriving economy centered on Baghdad, Mosul and Basra. This economy was based on the agricultural output of the hinterland, sustained by a sophisticated and well-engineered system of irrigation, employing huge stretches of canals and superbly constructed dams. This flourishing system was later destroyed beyond recovery by nearly two hundred years of neglect. There was a time, however, when Iraq had a thriving population of between twenty and thirty million and when Baghdad (the capital) may have had as many as two million inhabitants. These populations are almost twice as high as those of today and give some idea of the decline of Iraq since the time of the Abbasid caliphs.

The Iraq which fell under Ottoman (Turkish) rule in the sixteenth century became a remote and neglected province ruled from Constantinople (Istanbul). Until about one hundred years ago Iraq's cities remained isolated and backward city-states (*vilayets*), each

A water wheel still used for irrigation purposes on the Euphrates River.

controlled by an Ottoman governor with his garrison of troops, and each under the influence of a small group of leading city families. During the nineteenth century the pattern began to change, with the arrival of foreign consuls, traders and missionaries in larger numbers than ever before. Steamboats made their appearance on the Euphrates and Tigris rivers, and telegraph lines were constructed. After centuries of slumber Iraq began to stir once more. The Ottomans sent to Iraq new men of vision who tried to introduce reforms and changes which would modernize their sleepy

16

province; the beginnings of the Iraqi school system date from this time. Midhat Pasha, the famous Turkish governor of Baghdad from 1869 to 1872, was one of these well-meaning men whose reforms were unfortunately too little and too late.

On the collapse of the Turkish Empire, in 1918, the Ottomans left behind them an inadequate administration and a disorganized array of city notables and tribal chiefs through whom the country had effectively been governed. Seven hundred years had passed under foreign domination, neglect and misrule, the people were uneducated peasants and the economy was in complete disorder. Iraq was in no position to govern itself and the British, who had liberated Iraq from Ottoman rule and had seemed to promise so much for the country and its diverse peoples, became the new masters in a joint agreement known as the British Mandate. This was only to last until 1932 when Iraq finally achieved independence as a kingdom ruled by the Hashemite King Faisal. ("Hashemite" denotes a member of a family claiming common ancestry with the founder of Islam—the Prophet Muhammad.)

Modern Iraq no longer has a royal family and has little time for the past. It is a fast-developing, oil-rich country in a hurry to make progress and heal recent wounds. But to understand the nation it is necessary to look at its history and culture.

2

Mesopotamia Through the Ages

The ancient Sumerians lived in the flat and sometimes marshy land of southern Iraq (southern Babylonia) near the present-day confluence of the Tigris and Euphrates. At that time these rivers did not meet but flowed directly into the Arabian Gulf about one hundred miles (one hundred and sixty kilometers) north of the current shoreline. Sumerian civilization flourished from 4000-2000 B.C. and gave the world the earliest writing as well as providing the roots for the development of literature and numerous technological achievements.

The people of Sumer used the plentiful supply of reeds which grew along the riverbanks and canals (to make straw mats) together with the rich earth (from which they made mud bricks). By layering the mud bricks with the straw mats and coating this with bitumen or tar (which oozed naturally from the ground) they were able to build solid and waterproof city temples. Sometimes they used bricks which had been baked in a kiln to give a more resistant weatherproof outside skin. Often the temple base was topped by a stepped

A present-day Marsh Arab boy punting a boat loaded with reed mats.

pyramid, or ziggurat, which formed a series of platforms towering above the city. The walls of the temple were decorated with painted friezes or inlaid with shells, cones, carved limestone or baked clay reliefs. The inside of the temple shrine was simple—usually just an altar or offering table of baked mud, a collection of statues and sometimes a small hearth for offerings. Around the shrine were rooms for the temple priests as well as store houses for the various treasures offered up to the gods by the city.

Attached to each temple or palace in Sumer were "tablet houses" or schools where scribes were trained in writing cuneiform

19

inscription, using the ends of reeds to make wedge-shaped marks upon clay tablets. The writing on the earliest tablets is referred to as "pictographic" since each symbol stands for a whole word. The tablets which have been found by archaeologists record in precise detail the comings and goings of farmers, craftsmen and traders.

At the end of the third millennium B.C. the peaceable and prosperous Sumerian city-states were replaced by the empire of the powerful King Sargon of Akkad. However, the culture and traditions of Sumer were to last much longer since the Akkadians adopted much of their literature, liturgies and language. Some measure of the advanced civilization of the Sumerians can be gleaned by noting that it was only at this period in history (2200 B.C.) that people in Britain were building Stonehenge!

The ziggurat at Agarquf—one of the most impressive in Iraq.

Sargon came from Akkad, a broad region to the north of Sumer, with its capital at Kish (near the site of Babylon). Not content merely to rule this region, Sargon extended his empire throughout all Babylonia; westward into Syria and as far as Lebanon, northward into Guti (now Sulaimaniya) in Kurdistan.

Through a succession of minor kings and rulers the empire eventually declined and the mountain armies of Guti overran the land. The once-great kingdom of Ur was overthrown in 2006 B.C. Little is known about the confused period which immediately followed this decline. The turning-point in the history of the region lies with the emergence of the Babylonian Empire under Hammurabi which brought to an end the age of independent small city-states and ushered in the era of Babylonian dominance.

Babylonian civilization was dominant from about 2000 to 1600 B.C., although it was not until 1793 B.C. that Hammurabi (the sixth and greatest king of Babylon—"The Gate of God") came to the throne. Hammurabi is remembered as a strong and peace-loving ruler, though he did not hesitate to expand the boundaries of his empire and to elevate the Babylonian god, Marduk, to the position of the king of gods.

Hammurabi's fame stems, not from his conquests, but from his reputation as a great administrator and the codifier of laws derived from generations of previous Sumerian rulers. The structure and organization of his kingdom was such that the state was able to survive as a powerful and civilizing influence for over a thousand years. Hammurabi also gave particular attention to the construction, maintenance and repair of the canal system—for water was very much the life-blood of the state. Hammurabi's legal code, set out on a *stele* of black basalt found at Susa in 1901, contains three thousand lines of cuneiform script detailing nearly three hundred laws covering marriage, family, property, commerce, agriculture and slavery.

Although Hammurabi had made Babylon the center of the most powerful empire in the world, his family only managed to remain in power for a short period, during which they fought battles against new enemies—the Kassites (from the Zagros Mountains) and the Hittites (from the northwest).

The Kassites took power and ruled from Agarquf and Babylon. Then a new threat loomed from the northeast—the Assyrians. As early as 1900 B.C. their kingdom was emerging between the Tigris and the Kurdish mountains. It gradually grew stronger and more ambitious. In about 1500 B.C. the Kassites and Assyrians drew up a territorial treaty dividing Mesopotamia between them. This was not to last, for in 1150 B.C. the Assyrians dislodged the Kassite rulers from Babylon and the city fell once more into decline. The Assyrian Empire reached its height with the accession to the throne of Ashurnasirpal II (883-859 B.C.). After a series of daring military campaigns, the Assyrians extended their boundaries to include Babylonia, northern Arabia, Upper Egypt, Jordan, Israel, Syria and southern Turkey. And, by the seventh century B.C., they had established a tribute-paying empire which eclipsed even that of Babylon.

The ruins and ziggurat of Ashur. For one thousand years this city was the capital of the extensive and powerful Assyrian Empire.

Ashurnasipal II founded the royal city of Nimrud, and built a canal from the Zab River, as well as the magnificent Northwest Palace and the fortified buildings known as Fort Shalmaneser. The giant guardian figures of winged bulls at the palace were so heavy that they were left behind by the excavators (between 1949 and 1963) though most of the treasures may be seen today in the museums of Baghdad and London.

The capital of the Assyrian Empire had, for a thousand years, been at Ashur. King Sargon II moved it successively to Nimrud, Nineveh and then to the new city of Khorsabad (which was abandoned after his death in 705 B.C.). The magnificent relief carvings found in the royal palace there depict Sargon in his many campaigns of conquest in the west. In marked contrast, his son

23

Sennacherib is remembered as an administrator and as the builder and restorer of Nineveh, the new capital. Excavations at Nineveh (now Mosul) were begun in 1845. They have yielded over 23,000 clay tablets recording the history and culture of the Assyrians. Fortifications, waterworks and temples built by Sennacherib have all been unearthed, including the great palace whose walls were decorated with nearly two miles (three kilometers) of sculpted relief slabs.

The last great Assyrian king was Ashurbanipal. He will be remembered as a scholar-king who assembled the great clay tablet collection in the royal library at Nineveh, including the literary masterpieces *The Epic of Gilgamesh* and *The Epic of Creation* as well as valuable guides to the grammar and structure of Sumerian and Akkadian language and script. These have greatly helped modern researchers in interpreting ancient inscriptions.

The last great civilization of ancient Iraq was that of the Neo (or New) Babylonian period under Nabopolassar and his more famous son Nebuchadnezzar (602 - 562 B.C.).

Nebuchadnezzar set about planning a new imperial capital at Babylon. His new city was a great square with high inner and outer defensive walls in which there were eight gates, the main one being the Ishtar Gate which was dedicated to the god Marduk (or Baal). The walls of this gateway were decorated with a bas-relief of sculpted bricks and blue enamelled tiles depicting various animals (bulls, serpent-headed dragons and marching lions). Flowing through the center of the city was the Euphrates River. The king

The ruins of Nebuchadnezzar's palace in Babylon.

also built a new Southern Palace, the ruins of which are still standing. It had five great courtyards and contained the famous Hanging Gardens of Babylon—a series of terraced roof gardens described by Herodotus, the Greek historian, as one of the Seven Wonders of the Ancient World.

Nebuchadnezzar fought several campaigns against the Egyptian pharaohs but will be chiefly remembered for his capture and sacking of Jerusalem in 597 B.C. The Bible records how the people of Jerusalem were taken captive and led into slavery "by the waters of Babylon" (a reference to the complex network of canals which surrounded the city and which were maintained and repaired by slave workers).

A mosiac showing Alexander the Great in battle.

It was a foreign invader, Cyrus the Persian, who was to lay siege to Babylon in 538 B.C. The city had double protective walls and, for a time, it seemed that the armies of Cyrus would have to make an almost suicidal assault. Then Cyrus had a brilliant idea; he would divert the course of the Euphrates upstream so that it no longer flowed into the city. He instructed his troops to make their entry into the city along the now dry Euphrates channel.

Babylon passed from being the religious, cultural, administrative, commercial and military capital of a great and independent empire to being a provincial center of the Persian Empire, ruled by an appointed satrap or governor. After an attempted uprising, the Persian King Darius ensured that the city could never rise again as an independent force. He did this by pulling down and destroying the city walls.

After conquering Babylonia, the Persians pushed westward in pursuit of further territorial and material gains until they reached Greece. During this time the focus of events moved far from Mesopotamia to the great battles between the Greeks and Persians at Marathon (490 B.C.) and Platea (479 B.C.). The Greeks, under Philip of Macedon, fought back. Under his son, Alexander the Great, they succeeded in pushing the Persians eastward, defeating them in a great battle near Erbil in northern Iraq.

Alexander soon took the rest of Mesopotamia. He carried his campaign through Persia and on into India before returning along the Gulf to Babylon where he intended to site the capital for his great empire. Alexander set to work on rebuilding the temples and canals which had been allowed to fall into disrepair since Nebuchadnezzar's time. He embarked on a project to dredge the Euphrates to make it navigable between the Gulf and his new port of Babylon. But all these plans were to be of little use; in 323 B.C. Alexander, perhaps the greatest general the world has ever seen, contracted malaria and died of fever. His successor, the cavalry general Seleucis, then built a new Greek capital on the Tigris downstream of Baghdad. Seleucia replaced Babylon as a commercial and military center and soon grew to have a population of 600,000. Babylon reverted to an unimportant religious center devoted to the god Marduk.

The Seleucid dynasty ruled over the area for a further two centuries but during this period their territory was gradually whittled away by the advancing Romans in the west and the

Parthians from Persia in the east. The Parthians finally took control of the shrunken Seleucid Empire in about the year 200 B.C. and forced the Seleucids back into Syria. In 64 B.C. Syria became a Roman province and the Romans and Parthians became warring neighbors.

In four hundred years of power the Parthians never yielded to the might of Rome. Their dynasty was to be ended, however, by the Sassanian Persians. Under the Sassanid dynasty (A.D. 227 to 636) founded by Ardashir, the great civilization of Mesopotamia began to crumble. The ruins were to succumb to the silt of the river valleys and the sands of the desert plains.

The Sassanid ruler who is best remembered is Chosroes, the planner and engineer of the new city of Ctesiphon. Chosroes rebuilt the magnificent palace of Sapor with its great banqueting hall which was one hundred and twenty feet (thirty-seven meters) high, spanned by the Tak-i-Kisra arch (Arch of Chosroes)—the widest unsupported brick arch structure in the world. At Susa, Chosroes founded a university and a Persian school of poetry, philosophy and rhetoric. In addition, he arranged for the translation of the works of the great Greek and Indian writers, and he introduced the game of chess to the west.

The last Sassanian king, Chosroes II, completed the final chapter in the seven-hundred-year story of conflict between the Persians and the Romans. After defeat by the Romans, Chosroes abdicated the throne. The stage was now set for the invasion of Iraq by the followers of a hitherto unknown Arabian religious sect who would pursue a *jehad*, or Holy War, to have the world

Tak-i-Kisra—the Arch of Chosroes—the largest unsupported single-span arch in the world in the once magnificent palace of Sapor.

acknowledge the simple proposition that there was only one God, Allah, and that Muhammad was his prophet. This was the beginning of Islam.

Muhammad was a holy man from Mecca, now in modern-day Saudi Arabia, who claimed to have received a message from God through the Angel Gabriel. This message—the Koran—called for the submission of all people to a single God, Allah. The Islamic religion spread at a phenomenal rate after the death of the Prophet Muhammad on June 8, 632.

Muhammad left no sons to succeed him. Two prominent claimants to the succession emerged: Abu Bakr, Muhammad's

29

father-in-law; and Ali, his son-in-law and cousin. After some dispute, Abu Bakr was proclaimed Caliph (successor to the prophet), and he proceeded to lead an army on horseback north from Arabia to Iraq to attack and plunder the weak and unpopular Persian Sassanid Empire. Abu Bakr nominated his own successor, Omar. When he died it was the Caliph Omar who (with his great general Khalid ibn al Walid) was to lead the Arab forces into a series of wars of conquest.

The first great victory over the Persians was at Qadisiyah, in 637, where the Arabs stampeded the Persian elephants, killed their general, Rustam, and captured their battle standard of gem-studded panther skins. The armies of Islam proved unstoppable in the field. By 650, the empire of the Caliph stretched from Arabia in the south to the Caucasus Mountains and the Jazira and the Oxus Rivers in the north, and from the shores of the Mediterranean Sea in the west to the Indus River in the east. In the next century the empire would be spread even further westward to encompass all of North Africa and most of mainland Spain.

In Medina, the administrative capital of this fledgling empire, the Caliph Omar had meanwhile been murdered. A large section of the population was unhappy with his successor, Othman; he too was assassinated and the prophet's son-in-law, Ali, was declared Caliph. An opposition party, led by the Arabs in Damascus, refused to accept Ali and declared as Caliph the governor of Syria, named Muawiyah. It is from these rivalries to the rightful Caliphate that the two great subdivisions of the

Muslim world—the Sunni (Orthodox) and Shi'a (partisans of Ali)—arose.

Shi'a and Sunni tribal rivalry continued throughout a century of Umayyad rule from Damascus. By the middle of the eighth century the Shi'a rebellion against the Sunni had reached massive proportions. It had taken root in the east Persian city of Khorasan among Arab settlers and culminated in a drive westward by supporters of the Abbasid family (descendants and followers of the prophet's uncle Abbas). So began the golden age of Arab civilization—the Abbasid Caliphate.

The Abbasids moved their capital from Damascus to Baghdad where the Caliph Mansour built a magnificent new planned city called Medinat-as-Salaam ("The Abode of Peace"). The city was circular, more than two miles (three kilometers) across, and was surrounded by three concentric walls, each set with four gates. Great roads radiated from the palace in the heart of the city. The buildings from Ctesiphon, the Sassanid capital, were dismantled to provide construction materials for Mansour's city. Baghdad soon became the center of a blossoming culture and prosperous economy which was to last until the thirteenth century. A sophisticated system of taxation and expenditure on public works was instituted, along with an excellent road, courier and postal service; and, of course, the traditional and essential network of waterways and canals. Arabic became the official language and Islam the state religion but the Abbasids also retained and developed the very best of Persian literature, music, fashion, culture and architecture.

The Abbasid minaret at Ana on the Euphrates near the Syrian border.

The reign of the Caliph Harun al Rashid is traditionally regarded as the crown of Abbasid glory, since after this time (786-809) the eastern Muslim empire divided into independent kingdoms.

In 945 the Persian Buyid family were installed as rulers of Baghdad after a series of weak Caliphs. They devoted most of their efforts to extending the hold of the Shi'a faith in Iraq but they themselves were overthrown in 1055 by the advent of the Seljuk Turks who were strictly Sunni and who brought peace and prosperity to Baghdad once again. During this period the Caliph became more a figurehead than a ruler, since real power lay with the Seljuks. Several half-hearted attempts were made to restore the Caliphs to their former status, one of which was led by the Kurd Salah ad Din (better known to the Crusaders as Saladin), but the dynasty was in an irreversible decline.

The last noble deed of the Caliphs was performed by the Caliph Mustansir who, in 1229, successfully rallied his armies in a *jehad* (holy war) against the threat of invasion from Ogotai, the son of the Mongol Emperor Genghis Khan. But the Abbasid dynasty was doomed. The Mongol hordes returned thirty years later, in 1258, in the reign of the Caliph Mutasim to put to death 800,000 of Baghdad's inhabitants. Mutasim was executed and the Abbasid Caliphate was brought to an end. Mansour was destroyed and the prosperous population of merchants, traders, scholars and clergy were slaughtered. Iraq was to become no more than an insignificant province of a distant barbarous empire.

The invasion by the Mongols was repeated by Tamerlane in 1401 with equally horrific consequences for the people of Baghdad. The canal system of Mesopotamia was in part totally neglected and in part destroyed beyond repair. The Mongols deliberately drove the peasants from the sown land to create grazing grounds for their own flocks. After Tamerlane's death he was succeeded by Turkoman rulers from Eastern Persia.

Meanwhile, two other great empires were growing in strength beyond Iraq's borders. To the north in Turkey was Sultan Sulaiman's young Ottoman Empire; to the east the newly founded Persian Empire of the Safawid dynasty. The Persian ruler, Shah Ismail, was quick to act; in 1508 he took Baghdad. Ismail, a fanatical Shi'a, soon set about the conversion of the population, the destruction and desecration of the Sunni shrines in the capital and the persecution of the orthodox Muslims.

Having annexed Asia Minor, Sultan Sulaiman, a devout Sunni, sought revenge on his non-orthodox Shi'a neighbors. He marched south with his army of soldiers and artillery—it was the first time cannons had been used in Iraq; the Persian governor fled from Baghdad; the Sunni and non-Muslim population rose in revolt; and the Ottoman Sultan made a triumphant entrance into the city without any resistance. Sunni and Shi'a shrines alike were restored, so too were the canals and waterways; and the once-thriving commerce and trade of Baghdad was to some extent revived.

The Ottoman Turks appointed their own governors (known as Pashas) and divided Iraq into four administrative regions (the *vilayets* of Mosul, Kirkuk, Baghdad and Basra) and imposed strict public order through their garrisons of slave troops, called Janissaries. The Ottomans showed no great interest in regulating or assisting in the day-to-day affairs of the Iraqi people. Iraq had shrunk to the level of an obscure and neglected province of yet another empire.

During the eighteenth century the Ottoman capital, partly through lack of interest and partly due to the distractions of foreign wars, gradually lost direct control of the Pashas in Baghdad. Hassan Pasha, his son Ahmad and his successor, his

The Ottoman Turk Sultan Sulaiman—a contemporary portrait.

former slave named Sulaiman, founded a dynasty (the Mamelukes) which ruled for over a century and which finally broke the tradition of constantly changing Pashas appointed from Istanbul. In this period Iraq's commercial life experienced a long-awaited revival. Britain established a Trade Agency in Baghdad (a sub-office of the Basra branch of the East India Company), and for the first time European travelers began to visit Iraq, en route to Persia, India and the Far East.

In the early nineteenth century (following a dreadful plague which killed 100,000 Baghdadis), Sultan Mahmud II, the

35

energetic Ottoman reformer, deposed the Mameluke Pasha in Baghdad, abolished the Janissaries, imposed various administrative changes and reintroduced the direct rule of Iraq from Istanbul. It was not until the government of Midhat Pasha (1869-1872) that a complete reorganization of Iraq's administration was implemented. This enlightened statesman embarked on a new program of public works, road-building, sewage construction and educational provision. Iraqis were conscripted to the Ottoman army service and Turkish was established as the language of schools and government. Horse-drawn trolleys appeared on the streets of Baghdad. Municipal councils were established in the *vilayets*. But these reforms were too little and too late. The Ottomans lagged behind the progress being made simultaneously in Europe and in British India.

By the turn of the century Iraq had progressed little in four hundred years of Turkish rule—its people were largely uneducated and lacking in material goods. The economy of the country had not been developed; its agricultural and industrial resources lay untapped. It was in this climate that the roots of an Arab nationalist movement began to take hold. Ironically, it was the intervention of yet another imperial power, Great Britain, which was to rid Iraq of the Ottomans and pave the way for an independent Iraqi nation.

4

Twentieth-Century Iraq

At the outbreak of the First World War (in 1914), Turkey allied itself with Germany. Britain declared war on the Turks on November 7, 1914. Within a few days a combined force of British and Indian troops landed near Basra to seize the city and secure British oil interests in south Iraq. The following year saw the British advance in two armies up the courses of the Tigris and Euphrates. By the end of the war (1918) the Turks had been expelled from all areas of Iraq south of Mosul.

A decision about the future of Iraq was delayed until 1919. Then Britain declared that Iraq would be given limited self-government under a British mandate. This was a disappointment to many Iraqis who had hoped for complete independence, and it sparked off revolts in 1920.

As Iraq still lacked a natural leader, in December 1920 the British invited the Hashemite Faisal I to become king. Faisal came from a royal Arab family which had helped the British to drive the Turks out of Palestine and Syria. He had been crowned King of Syria in 1920. Then, when he was deposed by the French authorities, he had taken refuge in London. A referendum was held in Iraq and, on August 23, 1921, Faisal was crowned. During Faisal's popular twelve-year reign great progress was made in irrigation and road-building and the petroleum industry was extensively developed following the discovery of oil in 1927. In 1932 Iraq was admitted to the League of Nations as a sovereign independent state and the British mandate ended.

King Faisal died in 1933. He was succeeded by his son, Ghazi, who died in a car accident in 1939. During this period Iraq became more and more unstable. Between 1936 and 1941 there were seven attempts by different factions to seize power, the last of which was financed and assisted by the Germans who were unhappy that Iraq had sided with the British in the Second World War. In May 1941, British troops entered Iraq from Syria, drove out the rebels and re-established rule under the infant king Faisal II and his regent Abdul Illah. In 1953, the eighteen-year-old king assumed effective rule.

By this time Iraq had made enormous progress in social and economic development, as a result of wise spending of its oil revenues received from foreign petroleum companies for drilling concessions. And there was now a growing feeling within the country that perhaps Iraq should stop being so friendly with Western nations such as Britain, and follow the example of Egypt's President Nasser who was asserting Egypt's right to independence and encouraging Arab nationalism.

In 1955 King Faisal and Prime Minister Nuri al Said signed a joint defence treaty (the Baghdad Pact) with Turkey, and later with Britain, Iran and Pakistan. A special agreement was drawn up on

April 4, 1955 between Britain and Iraq to "maintain and develop peace and friendship between their two countries." The Iraqi government knew that it was becoming increasingly unpopular and that, without strong British support, the nationalists would overthrow the regime. It was not long before this happened.

In early 1958 the Hashemite kingdoms of Iraq and Jordan formed a federation in an attempt to prop up their respective royal regimes. On July 14, 1958, King Faisal of Iraq, the ex-regent Abdul Illah, and Prime Minister Nuri al Said were all assassinated and their mutilated bodies were dragged through the streets by a rejoicing crowd. The British Embassy was stormed by the mob and a staff member was killed. The British Information Office and Consulate were looted and burned.

The Hashemite King Faisal I who reigned for twelve years.

The original British Embassy in Baghdad, besieged and stormed by the mob during the 1958 revolution.

Brigadier Kassem, the leader of the uprising, became President and the Republic of Iraq was proclaimed.

Almost overnight all signs of the old regime were removed; the flag of unification with Jordan was replaced by the Iraqi flag; traditional place-names dating back to British rule or bearing royal connections were replaced; and all shops, schools and other buildings were decorated with posters of the revolutionary leaders.

However, peace and prosperity were not to be achieved overnight. Political support for Kassem soon evaporated. The Kurdish minority in northern Iraq rose in rebellion; so too did the Communists. Kassem withdrew Iraq from its Western alliances and renounced the Baghdad Pact. In 1961, Iraq unsuccessfully laid claim to neighboring Kuwait (a territorial claim which it has never renounced), and the regime found itself with very little international or internal support. Kassem was assassinated on February 8, 1963. His regime was deposed by a group of nationalist officers and Ba'ath Party supporters (including a young enthusiast named Saddam Hussein Takriti). Kassem's partner in the 1958 revolution was Colonel Abdul Salam Arif who became President. The Ba'athists (a pan-Arab Nationalist Socialist party founded by the Syrian Michael Aflaq) only lasted nine months in power; they were unpopular and internally divided and President Arif soon replaced them with more moderate army staff. In 1966 Arif died in a helicopter crash and was succeeded by his brother.

Throughout this period, Iraq's government could barely function; the country was in turmoil, uprisings were brutally suppressed, political opponents were imprisoned or executed, ministers were arrested and development projects neglected. Unsuccessful attempts were made to unite with Syria and Egypt. Iraq suffered economically from an oil pipeline dispute with Syria and from a 1967 decision to embargo oil supplies to the West in retaliation for their support of Israel in the Arab-Israeli War. On July 17, 1968 Arif was deposed by the Ba'athists, led by General Ahmed Hassan Bakr, and Iraq was subjected to another reign of political terror during which numerous opponents of the regime were executed, exiled or imprisoned.

However, at last Iraq had a stable government and was able to draw up and put into effect ambitious development plans. The constant thorn in its flesh was the Kurdish movement for self-government—a problem which refused to go away.

In 1972 Iraq nationalized the Iraqi Petroleum Company and took possession of all the assets of foreign-owned oil companies. A treaty of co-operation with the Soviet Union was also signed in that year. The quadrupling of the world price of oil in 1974 came as a godsend to Iraq—almost overnight the country became rich and powerful. Great progress began to be made in all fields. Schools and hospitals were built, road and sewage industrial projects were embarked upon and Iraq moved steadily into the front ranks of the developing nations.

In 1974 Iraq attempted to resolve the long-standing Kurdish problem by proclaiming a so-called "autonomous region" with its seat of government in Erbil. These reforms stopped short of the political aspirations of the Kurdish leaders, Jalal Talabani and Mustafa Al Barzani, and began a year of vicious fighting between Kurdish rebels and government troops.

In 1978 Ahmed Bakr was persuaded to step down as President in favor of Saddam Hussein Takriti. Saddam continued the development plans begun by his predecessor and further strengthened the country's internal security, eliminating every potential opponent and surrounding himself with trustworthy colleagues from his home town of Takrit. In 1979 Iraq made an unsuccessful attempt to unite with Syria. Meanwhile, over the border, the Shah of Iran was deposed and the fundamentalist Muslim leader Ayatollah Khomeini returned from exile in Paris (where he had fled after being expelled from Iraq by Saddam

Hussein). Khomeini's revolutionary government steered Iran away from modern Western technology and ideas, and towards a society based on traditional Islamic values and laws.

Over the centuries Iraq and Iran have shared a long history of rivalry, war and dispute, often over their common border. A dispute over navigation and territorial rights in the Shatt al Arab waterway—Iraq's only outlet to the sea—erupted into war in September 1980. For most of the next decade the war continued to be a drain on the Iraqi economy and many development projects were slowed down or abandoned. Neither side seemed prepared to put an end to the conflict on terms which were remotely acceptable to the other. The war dragged on until 1988, causing massive loss of life and the virtual crippling of Iraq's precious oil installations. The Persian Gulf War of 1991 caused even more widespread destruction in Iraq with bombing by international troops.

A battle scene during the Iran-Iraq war fought from 1980 to 1988.

4

Iraq's Geography

The plains of Mesopotamia have summer temperatures which are among the highest in the world. Temperatures in excess of 120 degrees Fahrenheit (50 degrees Celsius) are regularly reached in Baghdad. During July and August the thermometer remains constantly above 100 degrees Fahrenheit (38 degrees Celsius). Most Iraqis sleep outside on the flat roofs during the summer months. Traditional Iraqi houses are built with an underground room, or *sirdab*, in which the family can escape the heat of the day. There are really only two seasons in Iraq: summer and winter. Spring and autumn are both very short (perhaps less than one month each in the Baghdad area). Summer begins in May and lasts until October. Winter is surprisingly cold for a country of Iraq's latitude.

In the southern part of Iraq the annual rainfall is only about five inches (thirteen centimeters), all of which falls in the winter months. The rain turns the plains and deltas into a sticky, muddy quagmire. Most country roads become impassable, as do the pavements in the towns and cities.

Traditional Iraqi houses with flat roofs on which people sleep in the summer months, in the town of Rawa on the bank of the Euphrates.

Iraqi homes are large and airy. They become very cold and difficult to heat during winter. However, temperatures in most of the country rarely fall below freezing-point and frosts are rare. The winter is much more severe in Iraqi Kurdistan where, for several months, snow covers the mountains near the Turkish and Iranian borders. Records for the northern city of Mosul show that earlier this century the temperature remained below the freezing-point for nine consecutive days. The traveler who experiences summer temperatures in the "furnace" of Baghdad will find it hard to believe that ski resorts have been established less than half a day's drive away.

Iraqi thunderstorms can sometimes be severe and release downpours which quickly flood the drains and sewage systems in towns and cities. They may also raise the level of the rivers quite dramatically. In the past these storms have often produced disastrous floods which have burst canals, washed away roads, destroyed bridges and river banks, and inundated homes. Today, however, well-engineered flood control systems keep river levels in check and divert floodwaters into large man-made lakes, such as Lakes Habbaniya, Tharthar and Razzaza to the west of Baghdad.

The predominant winds in Iraq are the Shamal, a northwest wind which blows between June and October, and the southeasterly winter wind from the Gulf, the Sharqi. Dust storms are fairly common in western and central areas. They darken the sky to a reddish hue and leave a film of fine sand over everything. The dust penetrates the cracks around the windows and doors of every house. Many Iraqi housewives wash the floor and dust the furniture several times a day during the dusty season. "Dust-devils" are another common hot-weather phenomenon. They are rapidly-moving high spirals of dust which can be quite violent. Also associated with calm hot weather in Iraq is the mirage. During the mirage the atmosphere just above the very hot ground is disturbed by the movement of hot air. This distorts vision and, in particular, sometimes gives the appearance of water on the horizon. To the great disappointment of the hot and thirsty traveler, the ripples on the far-off pool turn out to be no more than hot air!

Except in the alpine regions of the north and the low-lying southern marshes, most of Iraq's plant-life has to contend with conditions of extreme drought. The desert areas of the west and southwest can develop a thick, attractive and varied plant cover in the spring. But this soon vanishes and the areas are barren wastes for the majority of the year. The natural vegetation in most of Iraq is a mixture of scattered, low-growing perennial shrubs which live throughout the year and can survive the blistering heat, and a collection of bright spring-flowering annuals. These burst in full color from the parched soil after it has been moistened by the winter rains. And they scatter their seeds before wilting under the summer heat.

Throughout the plains the camel thorn is common, so too are mugweed, boxthorn, caper, rock-roses, sedges and rough grasses. Daisies, buttercups and poppies grow in spring. Sage, flax, thyme and milkweed are found further north, as is the so-called thorn cushion. In the cities, date-palms, orange and pear trees, rubber plants and eucalyptus are common. Along the riverbanks and the marshlands grow thickets of tamarisk and bulrushes as well as a

Spiky desert vegetation among the ruins of Ashur.

few poplars and willows and the date-palm which is found everywhere in Iraq. The liquorice plant which grows up to six feet (two meters) high is often found near rivers. Liquorice is extracted from its underground stems (rhizomes) and is used as a flavoring, particularly for sweets. The northern Kurdish mountains were once covered with vast stands of oak. These have gradually disappeared over the years as they have been cut down and used for firewood and charcoal. Some areas are now little more than scrubland although others are the home of the pine and the maple. However, without an adequate forestry policy even these are under threat.

It might seem surprising that, despite the roasting heat and long dry season, Iraq is the home of a wide variety of animals and birds, both permanent and migratory. Among the most common mammals are deer, mountain goat, jackal, hyena, wild boar, brown bear, rabbit, bat, wildcat, fox and the curious date-palm squirrel. There are eighty-eight species of wild mammals in Iraq as well as ten domestic animals such as horses, oxen, water buffalo, sheep, goats and (of course) camels. Occasionally, a traveler may come across a herd of wild deer grazing in the southern desert or see a long straggling camel train trekking behind some Bedouin nomads as they cross the arid wastes. In the towns wild pi-dogs run wild and thousands of cats forage through the rubbish heaps. Most people in Iraq do not keep pets, though some do have cats or dogs, and a few even keep a gazelle in their garden.

The hoopoe or *hud-hud*, one of some 390 species of birds in Iraq.

There are some 390 species of birds in Iraq, though many of them are only short-term visitors who leave behind the European winter and fly to the warmth of the Iraqi sun. A variety of birds of prey live in the mountainous north; they include falcons, hawks, vultures and owls. In the southern marshes the marsh harrier can sometimes be seen. In the towns of Iraq pigeons have multiplied to the extent that they are now a pest. Crows and titmice also breed in large numbers and the occasional European robins can be seen in gardens. Storks are a fairly common sight in Iraq. They build their nests on good vantage points. If a stork decides to nest on an Iraqi house, this is taken as a sign of good luck. Nests are often seen on mosques, minarets, public buildings or historic ruins such as the Arch of Ctesiphon. In the great date-

49

The giant white pelican, found in the marsh areas of Iraq.

palm plantations and in the extensive orange and pomegranate orchards of south and central Iraq, the small singing nightingale and the large and colorful Arous al-Bustan (the "bride of the orchard") add an exotic flavor. Another lovely Iraqi bird is the hoopoe or hud-hud (King Solomon's bird).

In the low marshy lands of southern Iraq large numbers of swooping swallows can be seen; so too can the kingfisher, diving for fish in the warm pools. Geese, ducks, coot and other waterfowl prosper in the remote and humid marshes where the giant white pelican also makes its home. The pelican is hunted for its throat pouch—much in demand as a traditional drumskin. Its wingspan may be as much as ten feet (three meters).

Perhaps the most successful wildlife species of the desert areas are Iraq's reptiles. They have adapted with skill and ingenuity to their

50

natural surroundings. Pale sand lizards are frequent in the south; great smooth-skinned jumping lizards in the north. The harmless and timid gecko (or *Abu Bress* as he is known in Iraq) is a familiar sight in every house. Snakes are less common in Iraq and there are few poisonous varieties.

Iraq also has a rich river and stream wildlife including frogs (and tree-frogs), turtles, toads, catfish and pike. River fish from the Tigris, grilled in the open air on a roaring fire and served as *masgouf*, are a great Baghdad delicacy.

For thousands of years the Euphrates and the Tigris—Iraq's two great waterways—have served the country as its main highways. However, navigation in the middle and upper reaches of each river has never been easy, due to swift-flowing currents and a tendency to burst through the containing embankments. This seasonal flooding has had the secondary effect of washing away many of the roads along the river valleys and plains. Because of these difficulties in communication in earlier times, many minority groups and communities of differing cultures and ways of life survived relatively undisturbed through the centuries. The best example is perhaps the Marsh Arabs who live on boats, rafts and floating islands. Communications today are very much focused on the Iraqi capital, Baghdad. Even in the twentieth century, all major routes follow the course of the Tigris and Euphrates and their main tributaries, the Diyala and the Zab.

Neither the road nor the rail system in Iraq was well-developed even before the recent conflicts and each is under tremendous strain

51

as they try to cope with the heavy demands of a developing economy in a country which is virtually landlocked (has almost no access to the sea). The major towns are linked by poorly surfaced and poorly maintained highways, many of which are badly rutted by the heavy trucks which bring Iraq its imports from Jordan in the west and Turkey in the north. The busiest roads are from Baghdad through Ramadi to the Jordanian border, and south from the capital to Basra. Outside the major towns some roads are still dusty and pot-holed dirt tracks.

Iraq has ambitious plans to expand its road network. A highway was being constructed which would eventually link Basra (and Kuwait) to Baghdad and then continue into Jordan and the Red Sea port of Aqaba. This is particularly important for Iraq since the closure of the port of Basra during the long war with Iran has necessitated a great dependence on Jordan as an essential supply

A marsh Arab village built on rafts on the water.

route. A second highway will link Baghdad, Mosul and the Turkish border.

At first glance Iraq looks as if, like India, it is tailor-made to have a thriving railway network. Much of the country is a flat, virtually featureless plain and there are few steep gradients anywhere. But the present-day system is poorly planned and badly co-ordinated. It is seldom used for passenger traffic and is comparatively little used even for freight. Linkage between different sections of the lines is made very difficult because several different gauges have been used and the width of the railway track differs from region to region. The closure of the Iraqi-Syrian border has also meant that it is now no longer possible to travel by train from Iraq northwest to Europe. Iraq's railway development plan envisages the construction of some 1,550 miles (2,500 kilometers) of new track linking all the major cities within Iraq and joining up with the Jordanian, Turkish, Kuwaiti and Arabian Gulf systems (some of which are themselves still at the planning stage).

Iraq is linked in peacetime to most European and Arab capitals by air services, though these have been disrupted both by wars and by the government's ban on international travel by its citizens. Domestic air services exist between Baghdad, Mosul and Basra, where a new airport is under construction, although there has been little air traffic since the end of the Persian Gulf War.

5

The Cities of Iraq

Baghdad is a sprawling city of nearly five million inhabitants, divided by the great Tigris River. It has been the capital of Iraq since its foundation in the eighth century by the Abbasid Caliph Mansour. Baghdad is a place-name which for many people conjures up images of the Arabian Nights, Ali Baba and flying carpets. However, the modern city is far from romantic and retains virtually none of the traditional elements of the rich and mysterious east. Flood, decay, war and the city planners together have destroyed almost all parts of the city which are over fifty years old. With the exception of a scattering of old mosques and of two much-restored thirteenth-century structures (the Abbasid Palace and Mustansiriyah School), the city of Baghdad dates from the end of the First World War (1918) when its first two paved roads—Rashid Street and River Street—were constructed by German engineers to ease the passage of the Ottoman army in its retreat before the advancing British forces.

Since the end of the Second World War (1945) Iraq's new-found oil wealth has led to a rush to replace the old and the traditional

The modern city of Baghdad which bears little resemblance to the traditional romantic image of the so-called "mysterious east."

with the new, the modern and the fashionable. It is really only in the last twenty years that attractively designed buildings have begun to appear in the city. As a result of earlier poor and hurried planning, most of Baghdad is drab concrete, cement and unfaced brick, with poorly-maintained roads crammed with traffic of all types and lined by uneven pedestrian pavements obstructed with building rubble and piles of refuse.

To the credit of the city authorities, the numerous squatter settlements which developed around Baghdad in the 1940s and 1950s have been removed and the occupants have been resettled in new government housing developments in the outskirts of the capital. However, the government has been forced to check the drift of poor people from the country to the capital by the passing of special laws.

Baghdad used to be a city regularly stricken by plagues and prone to disastrous floods (the Tigris and Euphrates are within three miles–five kilometers–of one another) but a combination of investment in medical care, sewage systems and water plants and an ambitious series of engineering projects has dramatically improved the health of the capital and has made the periodic flooding a thing of the past.

Baghdad is very much the administrative and commercial hub of Iraq. It is a major communications center and has an international airport capable of serving most European and Middle Eastern capitals. Brick and cement manufacturing, food canning and packing, and oil-related developments gave the capital its pre-war industrial prosperity.

For the visitor, Baghdad still offers a number of first-class international hotels despite the wars. But because Iraqi Muslims do not allow non-believers to enter their places of worship some of the most interesting sights of Baghdad are not open to foreigners. These include the Shi'a shrines at Kadhimain and Adhamiya, the leaning minaret of the Sheikh Omar Mosque, the Sunni shrine of Abdul Qader Gailani, the thirteenth-century Sitt Khatun Tomb and the Caliphs' Mosque as well as the splendid Um al Tubul "Martyrs" Mosque. The last remaining medieval city gate, Bab al Wastani, is worth a visit, as is the old caravanserai at Khan Murjan (now a restaurant). The copper market, Souk al Safafir, is a place where traditional craftsmen can be seen heating, beating and decorating various metals in ancient or modern styles and where souvenirs can be bought.

Bab al Wastani—Baghdad's sole remaining ancient gate.

Lying on the Shatt al Arab waterway, forty-three miles (seventy kilometers) from the Arabian Gulf, Basra is Iraq's major port. Basra (the home port of the fictional hero Sinbad the Sailor) has for centuries enjoyed a great commercial importance.

The city of Basra is connected to Baghdad by both road and rail, and has been a prosperous center for the export and import of commodities, particularly dates, grains and wool. A pipeline connects a new oil exporting and refining terminal with the oilfields to the north, though these installations have suffered greatly in the wars with Iran and the international coalition. The Shatt al Arab has been effectively closed to navigation since the war with Iran began in 1980 and hundreds of rusting hulks of stranded merchant vessels and oil tankers have been trapped in the channel.

Basra is also famous for its traditional and historic houses, called

57

shanashils. Some of them were rescued from years of decay and were expertly renovated as part of the city's heritage program. The city is criss-crossed with a web of canals and waterways and this has led to Basra being called, rather fancifully, the "Venice of the East." The old bazaars in the Ashar district are attractive to the visitor and a pleasant afternoon can be spent wandering around the creeks, date-palm groves and backwaters of the Shatt in one of the many motorboats which ply for the limited tourist trade.

Northwest of Baghdad is the city of Mosul which, with a population of just under a million, is Iraq's second largest city. Mosul lies on the right bank of the Tigris. Across the river, facing it, lie the ruins of the once-great city of Nineveh, the capital of the powerful empire of Assyria. Formerly, the city owed its prosperity to the large-scale manufacture of high-quality cotton. Indeed, the name muslin is derived from the name of this city. Even today, weaving is still an important occupation, as is trade in wool, cotton, hides, wax, gum and nuts, although the oil industry has become the mainstay of the region's economy. Oil was discovered in 1927 at nearby Ain Zalah and, in 1934, a pipeline was built to the major refinery center at Baiji. Oil-related industries which have sprung up around Mosul include the manufacture (from bitumen) of asphalt for surfacing roads. The use of bitumen for making a form of asphalt goes back to the time of Nebuchadnezzar's Babylon and may be seen today in the surviving sections of its once-great Street of Processions.

Modern Mosul is an interesting town to explore. It has a greatly

58

varied population–Arabs, Kurds, Assyrians and Turcomans (the latter from eastern Persia)–and it has the highest proportion of Christians of any Iraqi city. Within the crumbling city walls, hidden in the winding back streets there are numerous interesting old churches and mosques, among which are the thirteenth-century Chamoun al Safa church and the Syrian Orthodox Ma Toma (St. Thomas) church which boasts the relics of St. Thomas. The oldest mosque, the Al Chabir (or Great Mosque) was built in 1172 and stands beside the surviving minaret (tower) of an earlier Umayyad mosque from A.D. 640 which looks as though it is just about to topple over. The tomb of the Prophet Jonah (*Nabi Younis* in Arabic), lies on the top of a "tel" or ancient mound overlooking the ruins of Nineveh. It is a place of pilgrimage for Muslims, traditionally revered as the prophet's burial-place. The Bash Tapia castle and the remnants of the thirteenth-century palace of Sultan Badrudden Lulu stand proudly above the Tigris on a surviving section of Mosul's city walls.

The old *souks* (markets) of Mosul are of great interest, offering a variety of local Kurdish arts and crafts as well as articles imported from the Turkish and Iranian parts of Kurdistan. Mosul has always been a crossroads on important trading routes between East and West and in recent times has been the funnel through which road traffic between Iraq and Europe passed. It also lies on the (currently inoperational) standard-gauge railway from Baghdad to the Syrian frontier (and so to Istanbul and Europe). Mosul is the base for visiting the ancient sites of Ashur, Khorsabad, Nineveh, Nimrud and

The impressive remains of Mosul's ancient walls, towering above the Tigris.

Hatra, whose excavations have revealed so much of the rich culture of Mesopotamian and Assyrian civilization.

The city of Erbil has been continuously settled since 3000 B.C. and was one of the chief centers of Assyria. It is therefore one of the oldest centers of population in the world. It lies about fifty miles (eighty kilometers) east of Mosul, in the center of a rich agricultural area, and has a population of half a million, most of whom are Kurds.

The focal point of the city is the high citadel or *qala* which for centuries has allowed the city to dominate the surrounding plain. The Babylonians called this city Arbah Illah—the "City of the Four Gods"—from which the modern name Erbil (or Arbil) is derived. Erbil was a regional center during the Ottoman period and the *qala* is full of lovely old Turkish buildings, some of which have been restored. Today, Erbil is in the heart of the Kurdish autonomous area and frequently the scene of political controversy.

Near Erbil is a well-built stone irrigation canal which is some 2,500 years old and which once brought water twelve miles (twenty kilometers) to the city from the Bastura Valley. A cuneiform inscription can still be seen in Bastura, reading: *I Sennacherib, King of Assyria, have dug three rivers in Khani mountains above Erbil, home of the venerated goddess Ishtar, and made their courses straight.* Other features of

The *qala*, or citadel, which dominates Erbil.

the city include the Mudhaffariyah minaret, dating from A.D. 1132, and the now-reconstructed fortress set into the rock above the ancient walls of the city. The surrounding area has a number of mountain resorts where the city-dwellers of Baghdad came in happier times to "get away from it all" in the green and pleasant hills of Slahudding, Shaqlawa, Rawanduz and Haj Omran (a ski resort).

Northeast of Baghdad is Sulaimaniyah, in a beautiful setting in the Kurdish mountains close to the Iranian border. The city is one of the chief cities of Iraqi Kurdistan. It is the central focus of Iraq's Kurdish nationalist movement and, until the 1980s, was a major center for trade with Iran. The town has a population of 100,000 and the local economy is based on the growing of tobacco, fruit and grains. Sulaimaniyah Governate (region) is one of Iraq's Kurdish autonomous areas and borders the mountainous Jebel Hamrin. It is a thinly populated agricultural region famed for its woods, forests and kind climate.

With a population of over 400,000, Kirkuk is the fourth largest city in Iraq and is an important economic center. Its wealth comes from the Baba Gurgur oilfields discovered near the city in 1927. Kirkuk lies north of Baghdad and is the provincial capital of the region of Ta'meem (Nationalization). The mosque in Kirkuk is said to contain the remains of the Prophet Daniel. The Bible reference to the "fiery furnace" is believed to relate to the continuous fires which can be seen in some of the oilfields around Kirkuk, caused by the spontaneous combustion of the volatile gases leaking from deep underground. During the Second World War these had to be smothered to prevent Kirkuk's oil-wells being an easy night target for enemy bombers.

Kirkuk is built on a *tepa,* or mound, containing remains of five thousand years of habitation. It is the largest city of the Kurds (Iraq's main minority group) and is part of the autonomous region which was created in 1974. The government is concerned that the Kurdish majority may one day pose a threat to the region's oil production and is making strenuous efforts to persuade non-Kurdish Iraqis to move to Kirkuk in order to dilute the Kurdish nationalist sentiment.

Najaf is the site of the Shi'a shrine of the murdered Ali (the cousin and son-in-law of the Prophet Muhammad). Like Kerbala, some forty-eight miles (seventy-eight kilometers) away, it is visited by a million Muslim pilgrims each year—many of them from India, Pakistan, Iran and Afghanistan. The original tomb of Ali was built a thousand years ago; the current mosque was built in the sixteenth century on the site. The mosque is richly decorated with bright tiles bearing Koranic inscriptions in the characteristic style of Islamic calligraphy, doors paneled in gold leaf, and mirrored walls and ceilings. Neon lights and straggling electric wires detract from the effect but in this respect Najaf is little different from any other Iraqi mosque.

Throughout the centuries it has been the ambition of devout Muslims to be buried at Najaf. As a result, minibuses dot the road carrying stretchers on their roof-racks (sometimes draped in the

Pilgrims outside the mosque at Najaf.

national flag) and bearing the families of the deceased towards the great cemeteries of the Wadi al Salaam (Vale of Peace).

Southwest of Baghdad, on the edge of the desert, is Kerbala–a sacred city of the Shi'a Arabs and the burial place of Hussein, grandson of Muhammad. Many of Kerbala's population are of Persian descent. Some people earn their living from making bricks and tiles bearing Koranic inscriptions, though the chief produce of the area is grains and dates. Kerbala's great mosques are the goal of

64

thousands of pilgrims from all over the Muslim world and its streets are crowded with the devout, many of whom sleep for days at a time in the city's alleys and *serais* (inns) to prolong their stay in this holy place.

The twin mosques of Kerbala face one another across the city's streets and squares which teem with pilgrims. The shrine of Abbas is less magnificent than the great shrine of Hussein though both are striking gold-domed, marble-tiled structures on the top of which fly the red silk martyrs' flags. These shrines are built on the site of the original tombs and are believed to contain some of the richest treasures in the world though the mosques themselves had been allowed to fall into some degree of disrepair over the last hundred years. Since the 1970s, however, the Iraqi government has granted large sums to renovate the shrines of Abbas and Hussein as well as the other tombs and shrines at Kadhimain (Baghdad), Samarra and Najaf. These are now being restored to their original splendor and did not suffer during the Iran-Iraq war since they were specifically excluded as targets by the Iranians (who are Shi'a Muslims).

5

Customs and Culture

Iraq observes the Muslim calendar. This is governed by the phases of the moon and means that their year is eleven days shorter than that of the western world. The Muslims number their years from the date of the flight of the Prophet Muhammad from Mecca to Medina—the Hejira—in A.D. 622. The most important events in the Muslim year are the Prophet's birthday, the First Muharram (Islamic New Year's Day), Ashura (the anniversary of the martyrdom of Ali) and the holy month of Ramadan, the dates of which vary from year to year. During Ramadan devout Muslims abstain completely from food, drink, perfume and cigarettes during daylight hours. During the night special religious services are held and families make the most of the opportunity to eat and drink. The elderly, the sick, travelers and pregnant women are exempt from this extremely taxing period of fasting. Iraqi offices usually work shorter hours during this month to make allowances for their tired and weak staff. The end of Ramadan is marked by a three-day holiday at Eid-al-Fetaar (the "Breaking of the Fast") when families celebrate with great parties and feasts and the exchange of gifts.

Other landmarks in the calendar are the Eid-al-Aadtha (the time of the annual pilgrimage to Mecca); Nawruz (the spring festival, celebrated particularly by the Kurds in the north); July 14 (the anniversary of the 1958 Revolution); July 17 (the anniversary of the 1968 Revolution); and February 8 (anniversary of the 1963 Revolution). In addition to the above, the Christian festivals of Easter and Christmas are observed by the substantial Christian minority. Friday is the Muslim holy day and is the weekly rest day for Iraqi workers. A six-day working week is the norm.

Iraqi names do not follow the Western pattern of Christian names and surnames. People are given a first name and customarily add their father's and grandfather's first names. So if someone is called Tahsin Ali Hassan, it is possible to deduce the names of his father and grandfather. Sometimes the family or tribal name is added; for example, Al Takriti ("inhabitant of the town of Takrit").

Although the Arabs have a tradition of hospitality and courtesy towards strangers, Iraq is not a country which receives many foreigners. It is perhaps a legacy of Iraq's colonial past through centuries of foreign domination that Iraqis should now be suspicious of strangers but also a result of the recent wars and tensions in this Middle East country.

They have, however, enthusiastically adopted Western clothes and fashions. In Baghdad officials wear smart suits, shirts and ties even in the hottest of weather. European-style dress is worn by most men and by many women, although women in rural areas still usually drape themselves completely in a black shroud, known as an *abayah*. Women tend to dress modestly and very often wear black.

67

A woman dressed in a traditional *abayah*.

Women's trousers, jeans and sleeveless dresses are not acceptable on the streets of Baghdad. Make-up is worn by most women and hair dyeing is very popular. The most coveted item of jewelry is a gold bracelet and every woman wears several–to some extent the prosperity of an Iraqi can be gauged from this clue alone.

The rural, less wealthy population are much more traditional in their dress. Men often wear the traditional *cheffiyeh* or checked headscarf–red/white or black/white–with its twisted black coil or *agal*, to keep it in place, and a *dishdasha* (a long gown) with sometimes a warm brown woollen cloak or *aba* over it. Women wear

68

the *abayah* or, particularly if working in the fields, colorful and voluminous peasant dresses. Some wear a gold-set jewel or *warda* in their noses and many have tribal tattoos. Women can regularly be seen in rural areas carrying loads on their heads and, even in the cities, it is not uncommon to see a housewife carrying her shopping-bag in this way. For casual wear, pajamas are popular among Iraqi men both indoors and out, and groups of pajama-clad Iraqis, some wearing robes, are a common sight on the street corners or around the coffee house (*gahwa*) as they spend their evenings chatting, smoking and playing dominoes. All Iraqi men have moustaches.

It is rare for men and women to socialize together in Iraq and

Two men on a ferry on the Euphrates, wearing traditional dress.

most houses have two sitting-rooms—one for the women and one for the men. Although many women work for a living, Iraqi society is still very much male-dominated. For instance, no Iraqi male would ever help with the household chores or the cooking, or let his wife drive the family car. It would hurt his pride too much to be seen in the passenger seat! This possibly stems from the fact that few Iraqis leave home before the age of twenty-five (and even then they often move into the house next door or build a special self-contained extension or *mushtamal*). They are therefore used to the constant attentions of their sisters, mother and grandmothers in their very large and close families.

Traditional Iraqi houses are large, usually brick- or mud-built, and have two large reception rooms. A dozen or more members of the same family may live there. Houses are usually two stories high and have a shaded courtyard in the middle which is open to the sky. Courtyard houses dating from 2500 B.C. have been found in the Diyala region of Iraq. There are narrow alleyways between the blocks in Iraqi neighborhoods, and bay windows often project above the alley. Many houses have habitable cellars which are cool but are prone to rising damp and surface water. At first-floor level the houses have an internal access gallery and vertically sliding sash windows. Roofs have a terrace with a parapet wall, which gives privacy when members of the family sleep there.

Meals are eaten communally and by hand from a large platter on the floor. Bread (*samoon*) is eaten with every meal. Few houses

Flat-roofed village houses made of mud brick.

have air-conditioning and most people sleep outside on the roof at night. Many keep televisions in their gardens or on their roofs. It is common for Iraqis (including children) to take a short rest in the afternoon and then stay up until midnight or later. People rise early—the noise of the cockerels and chickens which many people keep in their gardens acts as a general alarm clock, as does the dawn call to prayer from the *muezzin* at the top of the minarets of the city mosques. These calls are repeated five times daily.

Marriages in Iraq still tend to be arranged; often between cousins. The wedding is preceded by a long betrothal or engagement and the groom is required to pay a bride price to the girl's father (usually between five and ten thousand dinars). Muslim men are permitted to have more than one wife; and

71

polygamy and divorce both occur, particularly in rural areas. Iraqi society considers the birth of a male child an occasion for great feasting and celebrations. Men marry in their mid-to-late twenties while women tend to marry much younger. Children are born as soon as possible after marriage to prove the fertility of the father. The wedding itself is a great and costly affair which is preceded, even today, by a noisy cavalcade through the streets of the village or town.

Death is an occasion for great lamentation in Iraq. Women relatives of the deceased stay indoors for forty days, and dress in black during the year-long mourning period. Bodies are never cremated, nor are they buried in coffins. The deceased is washed, anointed with aromatic oils and then laid on a stretcher covered with an unstitched cotton shroud. The corpse is then placed in the grave, lying on its right side with its face looking toward Mecca (the Muslim holy city). Gravestones often carry photographs of the dead and bear small shrines in which some treasured personal effects are displayed.

Superstition still has a strong hold on modern Iraq. It is usual for a sheep to be sacrificed on a new car and at the gate of a new house to ward off evil spirits. Belief in the "evil eye" is widespread, and it is considered to be bad luck ever to compliment a mother on her children. Yellow eyes are considered unlucky but blue eyes are much desired.

Of all the Arab nations the Iraqis are probably the most enthusiastic about music, painting, and literature. Iraqi artists are celebrated throughout the Arab world and exhibitions of their work are held regularly in the various galleries in Baghdad. These are always well attended by the enthusiastic and knowledgeable public. Painting and sculpture are particularly popular, and many large attractive sculptures are on display in prominent outdoor positions throughout the main cities. Calligraphy is another art which has a long history of popularity in Muslim countries. This consists of ornate Arabic lettering, often recounting sections of the Koran, and is associated in particular with the decorations which adorn the walls of mosques.

Iraqi folk music is usually played on traditional instruments of which the *quanoun* or *santon* (a zither with up to one hundred strings), the *oud* (a stringed instrument which is plucked and resembles the European lute) and the *joza* and *rebaba* (Arab fiddles) are the most popular. On special occasions when there is something to celebrate Iraqis love to beat on small clay and skin drums, called *dumbuks*, while ululating loudly—making a sound halfway between a whoop and a yodel. Classical (Western) music is less popular, though Iraq does have a National Symphony Orchestra. The young people, like teenagers all over the world, often prefer popular music. Music is studied more seriously, though with just as much enjoyment, at the Baghdad School of Music and Ballet which was founded in 1970.

Although various types of writers and dramatists can be found in Iraq, it is poetry that is most popular with the public. Arabic poetry is written to be listened to, not just to be read. At an Iraqi poetry-reading the writer becomes a performer and the audience listen and applaud just as if they were attending a concert. Poetry is considered

73

Large sculptures like these, both ancient and more recent, can be seen throughout Iraq on city streets and in museums.

the greatest of the arts and Iraq's leading poets are internationally famous.

Iraqis really enjoy food. There is not a great variety among Iraqi dishes but the country's traditional food is both delicious and nutritious. Most popular are the ever-present *kebab* of lamb and tomato roasted on a skewer over an open fire; and *kubba*—spices, minced meat, nuts and raisins. *Masgouf* is traditional grilled fish from the Tigris usually cooked and eaten outdoors on the riverbank. *Pacha* is roasted sheep's head and is not for the squeamish! *Quzi* is a specialty dish usually served when twenty or thirty guests have come to dinner. It is a sheep, roasted whole and stuffed with rice, nuts, vegetables and spices. All these meals are eaten with great

74

quantities of *samoon* (flat unleavened bread) and washed down with *laban,* an unsweetened yoghurt drink. The meal will be rounded off with several cups of hot, thick and sickly-sweet Arab coffee (similar to Turkish or Greek coffee) or equally sweet lemon tea (*chai*) followed by a glass or two of *arak,* the local liqueur. This is a powerful distillation made from dates which is guaranteed to give the inexperienced drinker a terrible headache the next morning!

There are few countries which have a daily newspaper devoted entirely to sports; Iraq is such a country. Iraqis love sports—they love to play them, read about them, talk about them and watch them on television. The most popular sport in Iraq is soccer and there is a thriving national league whose matches are shown live on television each week.

At every street corner and on every patch of waste ground there is a crowd of schoolboys chasing a ball. The star players of the big European teams are just as much household names in Baghdad as in Paris or London. Basketball, handball, volleyball, weight-lifting and boxing are the other main sports which command a following among the Iraqi public. Facilities have a long way to go before they can match those of Europe and North America but Iraqi athletes have the enthusiasm and the official backing which makes eventual Olympic or World Cup success an achievable ambition.

6

The Peoples of Iraq

The Arab population of Iraq, although Muslim, is split between the Shi'a (fifty-five percent) and Sunni (forty-five percent) religious sects. Iraqis may share a common language, Arabic, and feel a common national sentiment but they have very mixed racial origins. This is not surprising in the light of the long history of Iraq's conquest and occupation by other peoples. The largest racial minority are the Kurds and the largest religious minority the Christians. There are, however, many other minorities including the ethnic Assyrians, Turcomans, Armenians, Persians and Jews, and the Sabean and Yazidi religious sects.

The Kurds (from the Persian *gurd* or "hero") are a strong and restless people who live in the mountainous north and northeast of Iraq, in and around the cities of Mosul, Erbil, Kirkuk and Sulaimaniya. Traditionally they have been semi-nomadic, pastoral tribesmen who rear sheep and goats in the hills and valleys of Iraq's border country between Turkey and Iran. The Kurds speak their own language (which is a form of Persian and

Kurdish children dressed for an Eid celebration in brightly-colored embroidered dresses. The Kurds live in the mountainous north.

has two dialects) and have a strong sense of their own culture, heritage and identity. Kurdish peoples are found in an area which is divided between Turkey, Iran, Iraq, Syria and Russia and which is sometimes called Kurdistan.

The Kurds have a unique national costume, not simply donned on ceremonial or religious occasions but genuinely their day-to-day garb. Their colorful, embroidered jackets which are short and sleeveless, their baggy trousers, traditional *khanjars* (curved daggers) and cummerbunds are all worn very much against the trend of the younger westernized Iraqis who tend to choose smart European suits or more informal jeans and casual shirts. The

77

Kurdish girls wear headscarves and bright ankle-length dresses, petticoats and shawls in eye-catching shades, with sashes at the waist. During pregnancy Kurdish women wear a *duabend*, an amulet inscribed with Koranic verses to ward off evil spirits. Most Kurds wear characteristic "platform" shoes—*qalik*—made of built-up leather. The men often wear skullcaps or turbans, but these are usually discarded in everyday dress; a knitted woollen cap, or *klaw*, is sometimes worn instead.

Despite the granting of limited autonomy to the Kurds in 1974, armed rebellion has continued. The area including Mosul, Erbil, Kirkuk and Sulaimaniya became a "safe haven" for Kurds after the Persian Gulf War in 1991. Despite military patrols, however, the area still sees frequent conflicts between competing Kurdish factions, Iraqi forces and Turkish troops pursuing rebels from Kurdish areas in Turkey. The Kurdish problem has been a thorn in the side of the Iraqi government since the Ottomans left Iraq some eighty years ago—it is a problem that shows no sign of going away.

The Marsh Arabs, or Ma'dan, are a tribal group of dark-skinned peoples who inhabit the large area of permanent and semi-permanent reed marsh and lagoons between Nasiriyah, Qurna and Amara, around the confluence of the Tigris and Euphrates Rivers. The way of life of the Marsh Arabs has changed little in thousands of years, since the waters of the Gulf receded and left behind this watery wilderness. It is only in the last forty years that these areas have been explored and mapped by European travelers and, even now, the Ma'dan remain

Marsh Arab houses built from reeds.

something of a mystery even to their fellow-countrymen.

The Ma'dan live in arched reed huts floored with reed matting. These dwellings have no heating, electricity or running water. The Marsh Arabs have few possessions—usually a few head of water buffalo, a gun and a bitumen-coated reed canoe (called a *mashuf* or *tarada*). They live by fishing and by hunting the occasional boar and the wild fowl which are abundant in the marshes. Traditionally, the fish are caught with a five-pronged spear thrown from bows of a *tarada*. However, modern methods of fishing with poisoned bait, usually shrimps thrown on the surface of the water, now threaten to displace the old ways. It is a

79

custom of the Marsh Arabs that fish should never be caught by nets.

The pride of each floating village of marsh people is the *mudhif*, or guesthouse, where any visitor is welcome to stay the night. These *mudhifs* are usually owned by the village *sheikh* (headman) and it is a point of honor that no visitor should ever be refused hospitality and that there should never be any payment for this generosity. The *mudhifs* are often constructed on a grander scale than the simple reed huts of the villages but are of the same basic design. Carpets and cushions are laid on the floor for the comfort of the guests (chairs are never used) and the male hosts will sit down cross-legged with the visitors to enjoy a sumptuous feast, perhaps followed by music and dancing. Meanwhile, in this male-dominated society, the women are elsewhere cooking, washing or milking the water buffaloes.

The people of the marshes may owe their origins to the bands of ex-slaves and outcasts who first sought refuge and anonymity among the reeds during a ninth-century rebellion. The modern world has recently intervened in the formerly quiet life of the Marsh Arabs. They follow the Shi'ite form of Islam and Iranian influence has grown in their region. A low-level revolt against the Baghdad government has been carried on since the end of the Persian Gulf War in 1991.

The Bedouin, or Bedu, are nomadic camel-breeding tribes which for centuries have migrated seasonally between western and southwestern Iraq, Saudi Arabia and Kuwait. A glance at a map of the area shows a large neutral zone between the borders of the three countries through which the various nomadic tribes travel freely. The winters are traditionally spent in more southerly well-vegetated areas and the summers in and around the oases on the fringes of Iraq's Western Desert, near the Euphrates.

The Bedouin have traditionally earned their living by breeding camels and trading them in the village markets. The camels are valuable as a source of meat, hide and hair, and as a means of transport. The way of life of these nomads is both poor and harsh. The Bedouin have therefore been under increasing pressure to abandon their traditional lifestyles and move to villages or towns, where they can practise farming as settled tribesmen. Many have chosen the greater security of village existence. The Bedouin are a proud people who have long

A family group. Despite attempts to integrate them, the marsh-dwellers still maintain their traditional culture and lifestyle.

A black tent—the home of a Bedouin family.

despised the way of life of the small farmers–*fellahin*–of the villages. Consequently, though the number of nomads has dropped, they are still to be found in modern Iraq. Travelers on the highway from Kuwait to Baghdad who look westward may see small encampments of these tribesmen among the dunes, sheltering beneath the blazing sun in their black tents, or perhaps hunting with their prized hawks. The sight of a Bedouin camel train strung out in single file across a bleak desert is one few visitors to Iraq will ever forget.

There are several other ethnic minorities in modern Iraq, though none of them is as numerous as the Kurds. The Assyrians were originally found in the Kurdish mountains and in and around

82

Mosul. Many more fled to Iraq from Turkey during the First World War. They have their own language (a dialect of Syriac) and are followers of the Christian Nestorian Church. Some Assyrians were unhappy when they were incorporated as citizens in the newly formed Republic of Iraq and, in 1933, there was considerable unrest which eventually led to hundreds of deaths in clashes with the Iraqi army.

Since that time, however, the Assyrians have integrated well into Iraqi society. Today they are accepted on equal terms. So too are the Armenians–Christian refugees from the Ottoman Turks. There is a large Armenian population in Baghdad and they form a close trading community, with their own church and their own language and script.

Around the northern city of Kirkuk and in an area southeast of Mosul live a large number of Turcomans. These are the racially pure descendants of the central Asian Mongol invaders, led by Tamerlane who ruled Iraq seven hundred years ago. In addition, a small number of people of Persian (Iranian) origin are to be found living in the Shi'a holy cities of Najaf, Kerbala, Samarra and Kadhimain (a suburb of Baghdad). The war with Iran encouraged a certain amount of emigration from this group during the 1980s and it is likely that their numbers will continue to decline. Their situation is similar to that of the Iraqi Jews who numbered around 100,000 in the 1940s but who have all but disappeared today due to emigration as a result of persecution.

The Yazidis (from the Persian *Yazdan*, meaning God) are a fascinating and religious people who live around Jebel Sinjar in northwest Iraq. They speak a Kurdish dialect but use Arabic in their worship and regard both the Bible and the Koran as holy books. They are sometimes called "devil worshippers" but this is not really a fair description, although they do pay tribute to Satan. They regard him as a fallen angel who will one day be reconciled with God but who, meanwhile, needs soothing gestures to keep him happy.

The Yazidis dress in white voluminous robes and red turbans. They worship at the sacred shrine at the tomb of their founder, Sheikh Adi, with its characteristic white-fluted steeples and its holy emblem: the Peacock Angel. At the shrine is the spring of Zem-Zem, which is believed to flow directly from Mecca and which is used for the baptism of children. Some other unusual aspects of their religion are that Yazidis will never step on the threshold of their shrines, and that they dislike the color blue and avoid eating lettuce, broad beans, pumpkins and radishes. Out of respect for the Prophet Jonah, they do not eat fish, nor do they eat the meat of male poultry, lest they offend the Sacred Peacock, Azreal. Yazidis also practise both snake charming and snake swallowing.

Another minority group in Iraq are the Sabeans (or Mandeans), sometimes called "the Christians of St. John" because they claim to be followers of John the Baptist. This is largely inaccurate since, with the exception of their custom of baptismal immersion in water, their beliefs are related to the Muslim faith. They are mentioned in the Koran as one of the "peoples of the book." Ritual washing plays an important part in their religion; and, when praying, they face the Pole Star.

The Sabeans originate from the banks of the rivers in southern Iraq (in an area around Amara, Qurna and Souk al Shiukh) where they are renowned as boat-builders though there is also a small community of them in Baghdad. The Baghdad Sabeans are bushy-bearded craftsmen, specializing in making silver bracelets, necklaces and rings decoratively inlaid with antimony (a silvery metallic element). These often depict characteristic scenes of Iraq, such as palm trees, camels, boats and reed dwellings. Sabeans are vegetarians and pacifists. In the past, they have been excused army service because of their religious requirement to live beside running water.

8

Education, Health and Society

One of the government's more successful development programs is education. Almost every village in Iraq has at least an elementary school and every town has a secondary school, although village schools are often built of mud brick and have tin roofs. All education is provided free by the state. There is no formal school-leaving age—since 1979 all children have to attend elementary school and about half of them continue to secondary level.

The origins of the Iraqi education system lie in the traditional Koranic schools—religious institutions attached to every mosque. For a thousand years these have provided a basic religious education in which the scriptures were learned "by heart" and the interpretations of the religious instructors (*mullahs*) were never questioned. Even today most students in Iraqi schools and colleges concentrate on memorizing facts; and they are taught to believe that the teacher is always right. The experimental approach is rarely used in the teaching of science. Examination

Iraqi children.

results and certificates are very important to the young Iraqi and a good degree or diploma is very much a passport to a good job and a good salary.

In 1850 it was estimated that fewer than one Iraqi person in every hundred could read and write. Fifty years later that figure had risen to about five percent. Today over fifty-eight percent of the population are literate and the government has mounted a major program of literacy training for all those between the ages of fifteen and forty-five who cannot read or write. A great deal of publicity has been given to this campaign.

Today Iraq has six universities (three in Baghdad and the others in Basra, Mosul and Erbil) and twenty-two institutes of technology, agriculture and administration. Education is given a high priority by the Iraqi government. They recognize that a

Iraqi doctors, dentists and pharmacists advertise their services in Kerbala.

literate and skilled population is essential if Iraq is to continue its social progress and its industrial and technological advancement.

During the 1980s, Iraq saw an enormous expansion in the provision of social services. Massive housing developments were completed, schools were opened in every village, and hospitals and clinics were set up throughout the country. In addition, a major health campaign with the slogan *Health for All by the Year 2000* was launched. Since 1968, the numbers of doctors and dentists in Iraq have doubled, and the number of pharmacists has increased fivefold. One problem which has yet to be overcome is the lack of qualified nurses, due largely to the very low status of nursing in Iraq. In the past, it has been considered a dirty and

88

demeaning job. In an attempt to remedy this position, the government made it compulsory in 1985 for all female school-leavers, college or university graduates to complete one year's nursing service before finding work elsewhere.

In September 1985 Iraq began the first stage of a national vaccination campaign which tried to protect all Iraqi schoolchildren against tetanus, diptheria, whooping cough, poliomyelitis, tuberculosis, measles and German measles. Improvements in flood control, water supply and sewage systems have gone a long way towards making the dreadful epidemics (in which thousands of people died) a thing of the past. In the great plague of 1831, for example, two-thirds of the population of Baghdad (100,000 people) died. In addition, government health education campaigns have improved public understanding of the importance of hygiene. Better housing conditions, provision of health care and changes in diet all contributed to falling levels of infant mortality and deaths among expectant mothers although some ground has been lost due to recent conflicts. To European eyes, however, most Iraqis appear to grow old before their time; and the average life expectancy, particularly in rural areas, is some ten years shorter than in the West. These problems are being tackled but there is still a long way to go.

9

Agriculture and Industry

Although agriculture in Iraq is considerably more developed than in any of its Middle Eastern neighbors, less than one-fourth of its land area is under cultivation. The remainder is desert, scrub, mountain, marsh and built-up area. Approximately twenty percent is cultivated farmland; twenty percent is pasture. The country's forest area is now less than 5 percent of the land because of excessive deforestation and unrestricted grazing. Agriculture employs about fifteen percent of the people who work.

Although the rainfall varies from region to region, and is generally low, the soil is often very fertile. Artificial irrigation is a long-established practice (dating back to Sumerian times) and is being extended in joint irrigation and flood control schemes. Although some crops are grown on mountain slopes in the north, the main crop-growing area is in the plains adjoining the Tigris and Euphrates. Here, flooding in the spring deposits a deep layer of rich silt on the valley floors.

The government is anxious to achieve a green revolution—to lessen Iraq's reliance on imported foodstuffs. Co-operative farms (worked jointly by a number of farmers) now account for some two-thirds of Iraq's cultivated land. About one-quarter of Iraq's agricultural land has been reclaimed—usually from salty areas which had fallen into disuse.

In modern Iraq large-scale irrigation projects are going ahead as a priority. The major new scheme is the Eski-Mosul Irrigation Project (near Mosul). Here, the waters from the dam will irrigate over one million acres (400,000 hectares). Other major projects are the Bakhma, Haditha (on the Euphrates near the Syrian border), and Dohuk Dams and the Hindiya and Faluja irrigation systems.

Thirty years ago farmers in Iraq were using techniques which were hundreds or even thousands of years old. The soil was tilled by hand or by animals, and crops were planted and reaped by hand. Now tractors, trucks and farm machinery have begun to take over. Wheat production, in particular, has expanded on the wide, flat and

Irrigated fields on the banks of the Euphrates River.

Cattle, feeding on reeds in the marshes.

easily cultivated plains. It is now the major crop. New road-building has helped get vegetable and fruit crops quickly to market. Food canning, refrigeration and freezing have helped improve the standards and freshness of produce. Greenhouses, fertilizers and insecticides have all contributed to a healthier, higher yield of all-year-round fruit and vegetables.

Iraq's principal crops are wheat, barley and rice, tomatoes, melons, dates, grapes, oranges, cucumbers and sugarcane. Meat, poultry, eggs, grain, rice, sugar and dairy products are all imported and are often in short supply.

The closure of the Shatt al Arab has put an end to Iraq's sea-fishing industry. However, a limited quantity of river fish is netted in the Tigris and Euphrates. Plans are in hand, however, to

92

develop a huge fishing-port complex near Basra.

The mainstay of the country's agriculture is the date-palm; indeed September 17th has been designated as Date-Palm Day. Eighty percent of the world's dates are grown in Iraq, mainly in the southern plantations around the confluence of the Tigris and Euphrates rivers. There are some thirty-three million date palms in Iraq. It is known as the "Tree of Life," and there are 531 different species. There are four main commercial types of date: Halawi, Khadrawi, Zahidi and Sayir. Various industries have grown up to take advantage of the abundant fruit—date syrup is produced for canning; date palm leaves are stripped from trees and used in matting or pulped for making paper; the wood of the tree is used for carpentry; and the rough fibres of the bark are made into ropes. Nothing is wasted.

The date-palm grows to a height of forty feet (thirteen meters). The dates grow in their hundreds in bunches beneath the spreading canopy of broad, needly palm leaves. It takes about eight years before a young date-palm begins to bear fruit, although some plants are harvested after only a few years for their succulent palm hearts. The mature date-palm lives for over a hundred years. Dates have been grown in Iraq for five thousand years and they are very much a staple of the country's diet. Dates can be picked at several stages of ripeness: Khalal (sweet and juicy), Ratab (moist and firm) and Tamar (sticky and toffee-like). They are eaten raw, roasted, ground or pressed into cakes. They are also distilled into a strong alcoholic spirit known as *arak*.

Most Iraqis have date-palms in their gardens and special date gardeners are a common sight in Baghdad as they go from house to house, shinning up the knobby tree trunks and hacking off the dead palms before fertilizing the flower clusters on the female trees. Later in the season they return to harvest the crop for the lucky householder.

For over sixty years the Iraqi economy has been dominated by the oil industry. Oil revenues represented something like ninety-eight percent of the country's earnings. It is a risky business to place all your eggs in one basket, so moves were made to diversify away from too great a reliance on oil. Until the 1970s Iraq had very few industries except for petroleum. The major industries were all in the Baghdad area and were concerned with electricity, water supply and construction. The only manufacturing industries were food- and drink-processing (especially dates and beer), cigarettes, textiles, furniture, shoes, jewelry and some small-scale chemical output. Over the last twenty-five years, Iraq has embarked on some major industrial developments, unrelated to petroleum, and has signed important trade agreements with other countries, notably France, Britain, Brazil, China, Korea, Bulgaria and West Germany, but most of these agreements were suspended after the Persian Gulf War.

An oil- and steel-works was built adjacent to the petrochemical complex at Khor al-Zubair near Basra and a steel mill was built with Soviet help in Baghdad. Work has begun to exploit Iraq's precious mineral resources of sulphates and phosphates. Since 1972 sulphur

Ripening dates in a date-palm grove.

has been mined near Mosul; and a phosphate-processing plant was constructed at Al-Qaim. The phosphates will be used at Basra to produce fertilizers for agriculture. Over a dozen cement-works have been built at Kerbala, Sulaimaniyah and Mosul. Other important industrial fields are pharmaceuticals, electrical goods, telephone cables, plastics, shoes, bricks and flour milling.

In response to steeply rising demand, the electrification of Iraq was begun. A network of oil-fired power-stations, transmission lines and sub-stations was constructed. By the end of the 1980s most of Iraq's rural areas had electricity and the country's generating capacity had risen significantly.

In the early hours of October 14, 1927, an event occurred which is perhaps the most significant in the modern history of Iraq. A worker in one of the exploratory boreholes near Kirkuk decided to inspect the diamond drilling-bit and carry out some maintenance.

When he unplugged the borehole he unleashed a sticky torrent of black oil and gas, gushing out to a height twenty-five times higher than a man. A new era had begun for Iraq; one in which oil was to be the keystone of the economy and the passport to economic development.

Iraq became a member of OPEC (the Organization of Petroleum Exporting Countries) and it benefited immensely from the decision in 1974 to quadruple the world price of oil. It came just at the time when Iraq was expanding its oil production. By 1980, Iraq was the world's second largest producer of oil (after Saudi Arabia). Very soon after the start of the war with Iran, the Iraqi oil terminals, refineries and pumping-stations near Basra on the Gulf were destroyed by Iranian bombing. After that Iraq was forced to rely on its northern oilfields and its pipeline across Syria and Turkey to the Mediterranean ports. Even these were soon disrupted—the line through Turkey was sabotaged and took some time to repair and the pipe through Syria was closed in April 1982 following a disagreement between the two countries.

Since then, the Turkish oil pipeline has been Iraq's only export route and oil production has dropped dramatically. This has meant that the country is no longer earning money from its main industry and has, as a result, stopped most imports of food and consumer goods. Many non-essential development projects have had to be delayed or cancelled because the country does not have the money to pay foreign contractors for the work. Meanwhile, Iraq has turned its attention to constructing alternative and safe pipelines for its oil exports. In July 1985 a new pipeline was opened to its then friendly neighbor Saudi Arabia; and plans were at an advanced stage for other lines to Turkey, the Red Sea and the Gulf of Suez. However, the diplomatic consequences of the Iran-Iraq and Persian Gulf wars stand in the way of Iraq reaping the full benefits of its oil resources. Iraq is still a major oil producer with current production of some 1.25 million barrels a day (compared with Iran's 2 million barrels) and another 59,000 million barrels of reserves still in the ground. Iraq could continue producing oil at current levels for over a hundred years if the estimates of petroleum geologists prove to be correct.

10

Prospects for the Future

Iraq is a country with a long and glorious heritage, rich in culture and steeped in history. Perhaps it is condemned to live forever in the shadow of its past achievements. By the mid-1980s Iraq had been reduced to seeking international aid. The country, desperately short of skilled labor and hard currency, was relying heavily on borrowed money to carry forward its own often grandiose development schemes. Iraq's Gulf neighbors, particularly Kuwait and Saudi Arabia, subsidized the war with Iran largely because they feared the consequences for their own Sunni Islamic nations of a victory over Iraq by the Iranians who militantly follow the Shi'ite form of Islam. In August 1988 a ceasefire in the Iran-Iraq war was negotiated by the United Nations. Despite the occasional clash the ceasefire has held up.

Shortly after Iraq ended this conflict, it became involved in another. Kuwait, whose independent existence had never been recognized by Iraq, pressed for repayment of the war loans and drilled for oil on land Iraq claimed as its own. In an attempt to end these on-going disputes with Kuwait, Iraqi president Saddam Hussein ordered the invasion of Kuwait on August 4, 1990.

Fearing that Saudi Arabia, which had also made extensive loans to Iraq, would be invaded next, international troops, led by the United States, were sent to guard the Saudi border. Eventually twenty-seven nations joined the coalition to protect Saudi Arabia and liberate Kuwait and on January 17, 1991 a massive air strike was launched against Iraq.

On February 24, international ground forces entered Kuwait and thousands of Iraqi soldiers surrendered. Iraq signed a ceasefire after four days of ground fighting. The ceasefire allowed for United Nation's weapons inspectors to make inspections of Iraq's weapon stockpiles but left Hussein's regime and military forces intact. Iraq repaired its roads, bridges, utilities and government facilities damaged in the extensive bombing surprisingly fast. The country remains impoverished by an international embargo which restricts oil sales and limits many types of imports. Iraq remains in a tense disagreement with the United Nations over the inspections agreed to in the ceasefire and these tensions could escalate if some agreement is not reached.

GLOSSARY

abayah   A black cloak worn by Muslim woman in the rural areas of Iraq

Allah   Muslim name for the one true God who is also the God of the Christian and Jewish religions

arak   An alcoholic drink made from dates

caravanserai   A resting place for caravans usually with an inn and open area for the caravan's animals

cuneiform   Ancient form of writing usually on wax tablets

jehad   A holy war for the spread of, or in defense of, the Muslim faith (also spelled *jihad*)

Mecca   The holy city of the Islamic religion located in Saudi Arabia

Mesopotamia   A common European name for the area that is now Iraq prior to 1918. It means "land between the rivers" in Greek.

Muslim   A follower of the Islamic religion and the teachings of the prophet Muhammad

sirdab   An underground room in traditional Iraqi houses, used to escape the heat in summer

souk   A Middle Eastern market place

vilayets   City-states

ziggurat   A step-sided pyramid built by ancient Mesopotamian civilizations

INDEX

Darius, King, 26
date palms, 8, 47, 48, 49-50, 93
Diyala River, 51, 70

E

East Africa, 14
economy, 15, 17, 31, 36, 38, 41, 42, 43,
    52, 62, 94, 98
education, 17, 28, 36, 42, 86-88
Egypt, Ancient, 14, 22, 25, 38, 41
Epic of Creation, The, 24
Epic of Gilgamesh, The, 24
Erbil, 7, 27, 42, 60-62, 76, 78
Euphrates River, 7, 9, 10, 13, 16, 24, 26,
    37, 51, 56, 78, 90, 91, 92, 93
Europe, Europeans, 35, 36, 37, 38, 41,
    42, 43, 53, 56, 59, 67, 73, 77, 78, 89,
    94

F

Faisal I, King, 11, 17, 37
Faisal II, King, 38, 39
"Fertile Crescent," 13
festivals, 66-67
First World War, 37, 54, 83
fish, fishing, 79, 84, 92
flooding, 46, 51, 56, 90
food, 70, 74-75, 90, 92, 93

G

Genghis Khan, 33
Germans, Germany, 37, 38
Ghazi, King, 38
government, 8, 39, 41, 42, 55, 65, 78,
    80, 87, 90
Greeks, Ancient, 14, 15, 27
Guti, 21

H

Habbaniya, Lake, 46
Hammurabi, King of Babylon, 9, 21

Harun al Rashid, Caliph, 32
Hashemite, 39
Hatra, 60
Herodotus, 25
Hittites, 22
housing, 44-45, 55, 70-71, 79, 88, 89

I

independence, 17, 36, 37, 38
India, 14, 35, 53, 63
industry, 36, 38, 56, 58, 94-97
Iran, 10, 13, 38, 43, 62, 63, 76, 77
Iran-Iraq War, 11, 43, 57, 65, 83, 96, 97,
    98
Iraqi Petroleum Company, 42
irrigation, 15, 38, 61, 90, 91
Islam, 15, 29, 30, 31, 43, 63, 80
Israel, 22, 41

J

Janissaries, 34
Jebel Hamrin, 62
Jebel Sinjar, 83
jehad, 28, 33
Jordan, 13, 22, 39, 52

K

Kassem, President, 11, 40, 41
Kassites, 22
Kerbala, 63, 64, 65, 83, 95
Khalid ibn al Walid, 30
Khomeini, Ayatollah, 42-43
Khorsabad, 59
Kirkuk, 7, 34, 62, 63, 76, 78, 83, 95
Kish, 21
Koran, 29, 63, 73, 84
Kurdistan, Kurds, 11, 21, 22, 40, 42, 45
Kuwait, 11, 13, 41, 52, 53, 80, 82, 98, 99

L

language, 14, 24, 36

Saudi Arabia, 11, 13, 80, 96, 97, 98, 99
Second World War, 38, 54, 63
Seleucia, 27
Seleucid dynasty, 27, 28
Seljuk dynasty, 10, 32, 33
Sennacherib, King, 24, 61
Shah Ismail, 34
Shamal, 46
Sharqui, 46
Shatt al Arab, 43, 57, 58, 92
Shi'a Muslims, 7, 12, 31, 32, 34, 56, 63, 64, 76, 80, 83, 98
souks, 59
Soviet Union, 42
sports, 75
Sulaimaniyah, 21, 62, 76, 78, 95
Sulaiman, Sultan, 33, 34
Sumer, Sumerians, 9, 13, 14, 15, 18, 20, 21, 24, 90
Sunni Muslims, 7, 31, 32, 34, 56, 76, 98
superstition, 72
Susa, 22, 28
Syria, 11, 13, 14, 21, 22, 28, 37, 38, 41, 42, 53, 59, 77, 91, 96

T
Takrit, 42
Takriti, Saddam Hussein, 11, 12, 41, 42, 43, 99
Talabani, Jalal, 42
Ta'meem, 62
Tamerlane, 33

temples, 18, 19, 27
Tharthar, Lake, 46
Tigris River, 7, 10, 13, 16, 22, 37, 51, 54, 56, 58, 59, 78, 90, 92, 93
trade, 34, 35, 57, 58, 59, 62, 81
transport, 31, 36, 38, 44, 46, 51, 52, 53, 57, 59, 92
Turcomans, 59, 76, 83
Turkey, 13, 14, 22, 37, 38, 52, 53, 76, 77, 96, 97

U
Umayyad dynasty, 10, 31, 59
United Nations, 12, 98, 99
Ur, 21

V
vilayets, 15, 34, 36

W
Wadi al Salaam, 64
wildlife, 24, 30, 48-51
women, status of, 71-72

Y
Yazidis, 76, 83-84
Yom Kippur War, 11

Z
Zab River, 23, 51
ziggurats, 19